Pacific Northwest
SCULPINS

Gregory C. Jensen

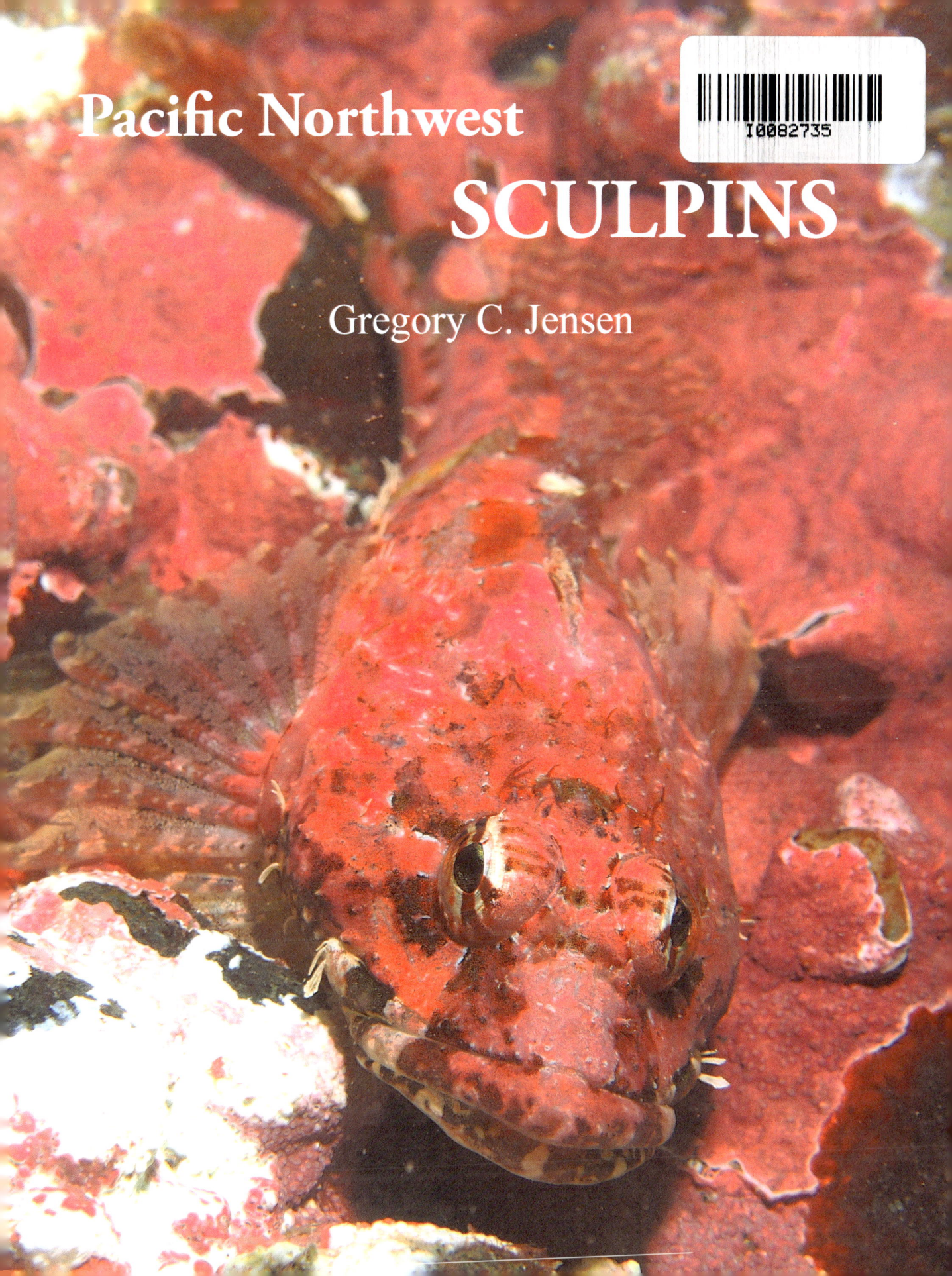

Adult male scalyhead sculpin (*Artedius harringtoni*)
in a giant barnacle shell

All photos © Gregory C. Jensen
            2025
ISBN 978-0-9898391-4-3

MolaMarine

Front cover: Buffalo sculpin (*Enophrys bison*)
Title page: Smoothead sculpin (*Artedius lateralis*)
Back cover: Staghorn sculpin (*Leptocottus armatus*) eating a
tubesnout (*Aulorhynchus flavidus*)

A pdf of this book can be downloaded for free at **www.molamarine.com**

Table of Contents

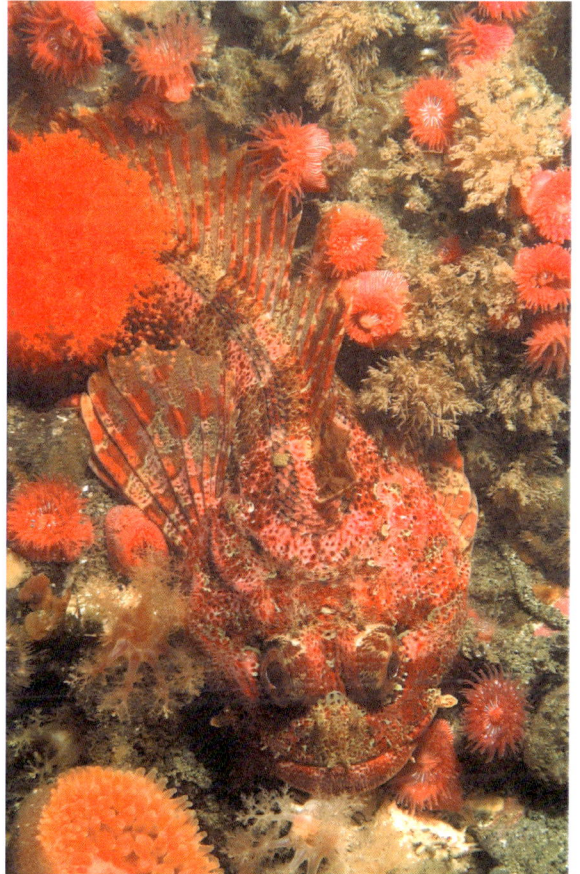

Red Irish lord, taking the "red" part of its name *very* seriously

Introduction

Underwater photographers in the Pacific Northwest have a love/hate relationship with sculpins. Photographers love the stunning colors and patterns that many species can have, and few groups of fishes are more cooperative when it comes to posing for the camera. But many divers get frustrated when they try to put a name to the sculpins they encounter, as the very color variations that make them such great subjects can also create a lot of confusion. Some species do have fairly consistent color patterns that come in handy, while many others change to match their backgrounds or moods. While the following guide will hopefully be of some help when it comes time to put names on your images, there are bound to be some that will leave you stumped due to the angle or lighting of the image. That's why if there is any question in your mind about the

Scalyhead sculpin

identity of your subject, try to take shots from several different viewpoint so that you'll hopefully capture some key features. And remember, there's no shame in putting "unidentified sculpin" on your image or on your REEF form. We all get ones that stump us.

This little guide is not yet comprehensive. There are still some species I have not encountered, or have only seen when I wasn't carrying a camera. It *does* include just about all the ones you're likely to see without making a special effort and/or being extraordinarily lucky.

"Sculpin" as used here means any fish that is, or used to be, in the Family Cottidae. Recent classifications place grunt sculpins in their own family, Rhamphocottidae, sailfin and silverspotteds in Hemitripteridae, and tadpole, soft, spinyhead and blackfin sculpins in Psychrolutidae. These are all grouped in the "Too Cool for Cottidae" section. The rest I've divided into four other groups: the "Poacher Wannabes" (sculpins that are often mistaken for poachers), with the remainder split up roughly by maximum size. The smallest are species less than 6 inches in length; the "Middle Class" are those 6-12 inches, and the "Big 'uns" that can grow to more than 12 inches. Some of these have also been recently reassigned to different families (e.g., cabezon and longfin sculpins as Jordaniidae).

LSD's: Little Saddlebacked Dudes, or, Sweating the Small Stuff

Most of the problem sculpins are little ones: species that max out at less than six inches, and often much less. And naturally, that takes in the majority of our local species. When you add to that the fact that all the big species have to start out small, things get even more complicated.

Just as birdwatchers have their difficult-to-identify "LBJ's" (Little Brown Jobs), fishwatcher's have their "LSD's" (Little Saddlebacked Dudes). These little guys have distinctive saddle markings along their backs. Several of the species tend to be overlooked by scuba divers because they're generally in water only a couple feet deep or less, like this tidepool sculpin.

Fluffy Sculpin, *Oligocottus snyderi*

This LSD is the only one with a line of cirri clusters running along the base of the dorsal fins; it also has a second, shorter row midway between the top row and the usual ones along the lateral line (though these tend to be multiple rather than single cirri). The dorsal fin spines are also tipped with cirri. They are often bright green to match seagrass or green algae, while those living in coralline algae can be pink. I occasionally find them in Puget Sound, but they are more typical of the San Juan Islands and outer coast. I've noticed that when caught, fluffies tend to keep their first dorsal fin erect while tidepools do not. Max length 90 mm (3.5 in).

The cirri can be hard to see, depending on the lighting and the color of the fish. This species also has a cirrus at the base of the nasal spine that is not present in the other LSDs, except for sharpnose sculpins .

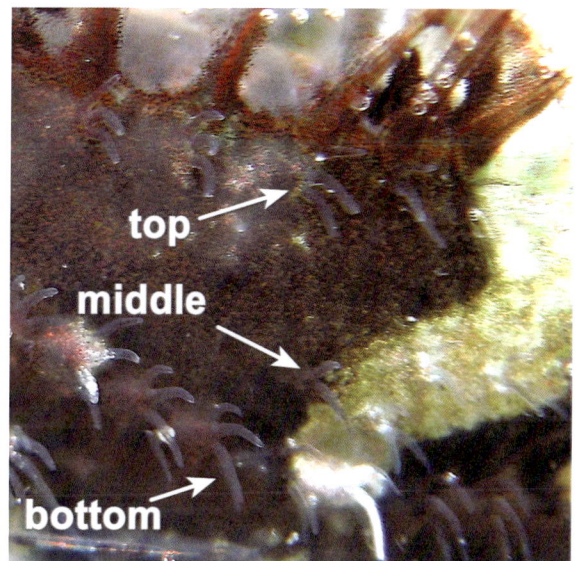

Fluffies can have beautiful markings on the underside

Tidepool Sculpin, *Oligocottus maculosus*

This is probably the most common LSD in our area, and the one that occurs in the shallowest water. At high tide, look for them hopping about among the barnacles, mussels, and rockweed that is found on the upper part of the shore. They vary in color from gray to green to nearly all black.

These fish can remember the location of their home pool, and find their way back from as far as 100 m (328 ft) away.

Tidepool sculpins have a two-pointed preopercle, but unless you carry a magnifying glass or have superhuman vision, you're not likely to see it in the field. They also lack scales, but it's nearly impossible to see the scales on the similar species that have them.

Note instead the size and position of the first white saddle. The front edge is at the middle or just beyond the middle of the first dorsal fin, and the rear edge is where the second dorsal fin starts. This saddle also extends well down the sides, past the lateral line. There is a line of cirri along the lateral line.

Maximum length is 90 mm (3.5 in), but typically about half of that.

Notice how well this tidepool sculpin blends in with the rocks and barnacles. The bold saddle markings break up its outline, making it very easy to overlook as long as it doesn't move.

Saddleback Sculpin, *Oligocottus rimensis*

This small intertidal sculpin can be found mixed in with tidepool sculpins. The preopercle has just a single point and, unlike the other LSD's, the body is covered with scales (small prickles, actually). Again, these features aren't going to be visible without a hand lens. The first saddle is neither a "V" like in the sharpnose, nor a broad band like that of a tidepool sculpin. The body is noticeably more elongate than that of either a tidepool or sharpnose sculpin. Maximum length 65 mm (2.5 in).

Sharpnose Sculpin, *Clinocottus acuticeps*

Sharpnose sculpins are often mistaken for tidepool sculpins, and the two species occur together in the intertidal zone. Sharpnose tend to have a light-colored area under their eye (sort of a reverse five-o'clock shadow). The first white saddle is in front of the first dorsal fin and the preopercle has a single point. Sharpnoses have cirri on the nasal spine and atop the eye; tidepools do not. Sharpnoses have a cirrus on the maxillary that is absent in fluffies.

The bottom right photo shows four sharpnose sculpins caught in the same place at the same time, showing the variation in their color and pattern. Note the forward-pointing "V" shape of the first white saddle (when present) and compare that with the tidepool sculpin.

Maximum length 63 mm (2.5 in).

Sharpnose sculpins are a tiny species- these four adults are in a small custard dish.

Spinynose Sculpin, *Asemichthys taylori*

This small, slender sculpin is generally found on shell hash or gravel bottoms in the shallow subtidal. It's a particularly interesting species for two reasons: it lays its eggs among the egg masses of buffalo sculpins, letting them do the guarding, and it is a specialized predator on small snails. It uses teeth in the roof of its mouth to puncture the snail shell so that the meat within can be digested. Some snails accidently get swallowed without being punctured, and can pass through the digestive tract alive.

This LSD can be distinguished by its elongate body with what looks like a zipper running down the side. It often has a "five o'clock shadow"- dark coloration below the eye. Color varies from black and white to coralline algae pink, and there is usually some bright blue edging along some of the saddles. There is normally one diagonal bar through the eye and the body has four dark saddles instead of the five or more in other LSDs. Maximum length 74 mm (2.9 in).

A pair of spinynose sculpins on the "classic" type of gravel/shell bottom where they are typically found.

Note the overall body shape, with a very flat outline from head to tail. When viewed from above the body narrows abruptly right before the tail, instead of gradually as in other LSDs.

Between their shape and the way they tend to drape themselves across the bottom, they often strike me as being very lizard-like.

Some authors put this species in the genus *Radulinus*, the same as a slim sculpin.

These fish can bury in rather course substrates

Northern Sculpin, *Icelinus borealis*

Northerns are one of the more challenging species to identify underwater. Besides the fact that they tend to live in silty areas and blend in well, there's no single easily seen and distinctive characters to focus on. Look closely at the head for: a large pair of cirri right behind the eyes (usually branched, note photo below); small cirri at the nasal spines; a cirrus at the corner of the jaw, and cirri on the tips of the first dorsal fin spines (bottom middle). They generally have several saddle markings. Could easily be confused with the Puget Sound sculpin, *Ruscarius meanyi*, in which the cluster of cirri behind the eyes consists of three separate fingers.

Maximum length 101 mm (4 in).

Manacled sculpin, *Synchirus gilli*

These little guys know how to be anywhere but in your plane of focus. They are very un-sculpinlike in their behavior, flitting about from one kelp blade to another. The pelvic fins are modified into a sort of suction cup that they use to attach to kelp, rocks, and pilings. With their long, slender body, pointed snout, and unique behavior they are not likely to be confused with any other local sculpin. Though they are often a drab kelp color, some have beautiful white markings. Maximum length 69 mm (2.7 in).

Although I see them occasionally in bull kelp in Puget Sound, they seem to be much more common in the Strait of Juan de Fuca where the constant surge adds to the challenge of focusing on them.

Introduction to *Artedius*

This trio of troublemakers probably causes more grief for fishwatchers than any other fish in the Pacific Northwest, since all three are common and very similar in appearance. In addition, scalyhead, smooth-head, and padded sculpins all show a lot of variation in color, pattern, and number of cirri; this can also all be complicated by changes with age and sex. There's also a possibility of undescribed, cryptic species in the mix that further muddy the waters.

The members of this genus have rows of scales between the dorsal fins and the lateral line that differ in width and scale number and size between the species. This, and differences in the preopercular spines are often used as distinguishing characters in books, as these are readily seen in preserved specimens in the lab (as seen in the padded sculpin below).

Unfortunately, these characters are hard to make out underwater and are often impossible to see even in a closeup photo if the lighting doesn't hit them just right. Soft, flexible features like cirri (hairlike projections) and nostrils are much more useful when viewing these fish in their natural habitat. These head features are shown and compared on the next page.

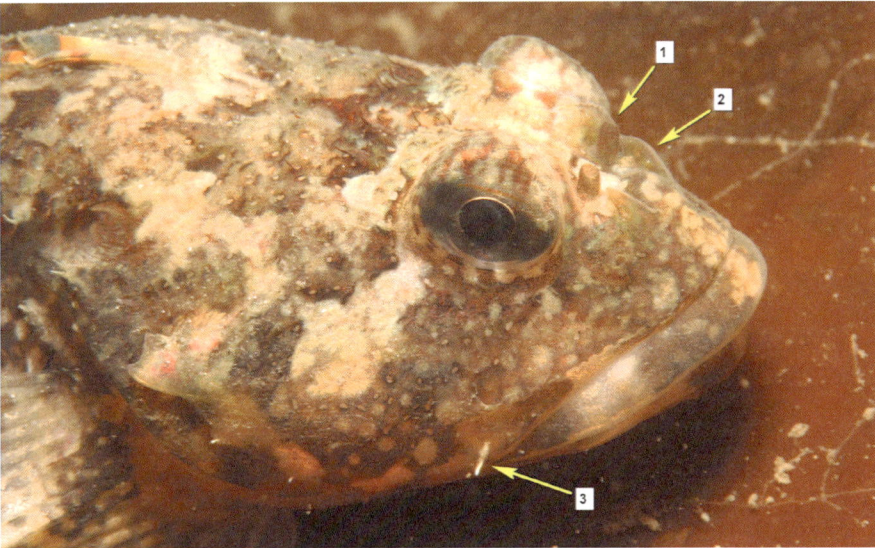

Padded

1. Large pair of nostrils near eyes

2. Large nasal spines

3. Single threadlike cirrus at corner of mouth

Smoothhead

1. Nasal spines and nostrils hard to detect

2. Longer snout

3. Usually three cirri at corner of mouth; may be flattened

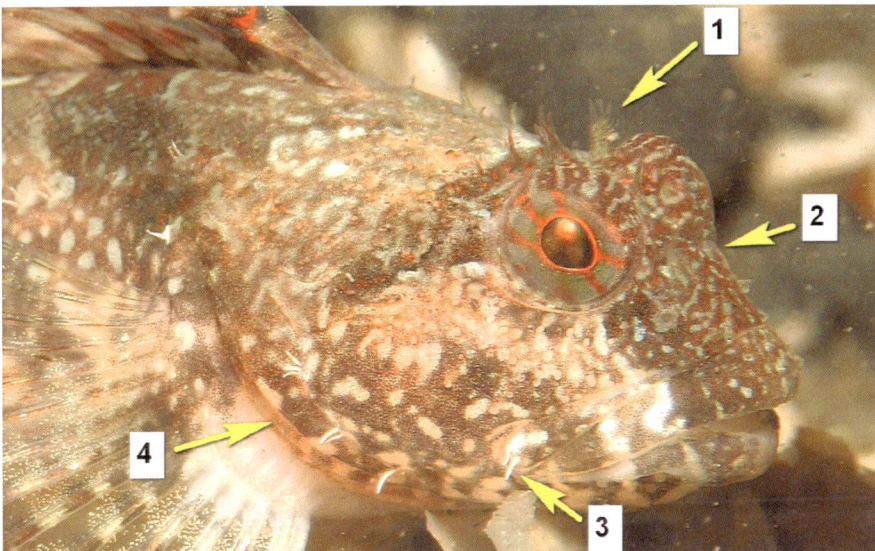

Scalyhead

1. Branched cirrus behind eye

2. Nasal spines and nostrils not obvious

3. Usually two threadlike cirri at corner of mouth

4. Gill membranes orange

Scalyhead sculpin, *Artedius harringtoni*

This is the species most commonly encountered by divers, in part because it might be the most common of the three, but perhaps more importantly because it occurs on reefs where divers spend most of their time. The above photo shows an adult male, which is the easiest of all to identify. Male scalyheads are the only ones that have large branched cirri between the front part of the eyes. The orange color under the gills is another useful character for this species.

This specimen shows several other scalyhead features, none of which are unique to scalyheads but in combination serve to identify them. The colorful radial 'spokes' in the eye, the bright white spot at the base of the tail, and the two threadlike cirri at the corner of the mouth all are characters suggestive of (but not exclusive to) a scalyhead.

The scales tend to be proportionately smaller in scalyheads than in the other two species; the width of the band is greater than in smoothhead and narrower than in padded sculpins.

Maximum length 102 mm (4 in).

Small juvenile scaly-heads (right)have a tiny, bright blue dot on the top of their heads.

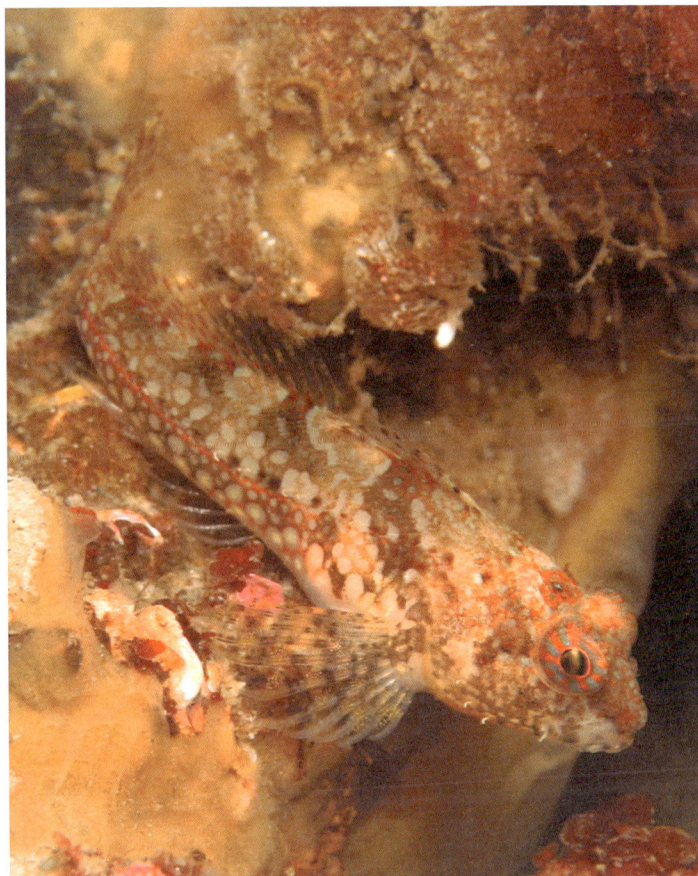

Big males have huge mouths and large teeth, in addition to the tall cirri between the eyes.

Although all three *Artedius* can have round spots below the lateral line, scalyheads tend to have the roundest, best-defined ones (left).

Don't expect all scalyheads to be brightly colored. Those that live in areas without coralline algae and other colorful organisms can be rather dull in color. Even those in very colorful areas like Deception Pass can be quite drab. Scalyheads like to hang out in empty barnacles (below).

Smoothhead sculpin, *Artedius lateralis*

Smoothhead sculpins tend to be more drab than scalyheads, but as the specimen above shows there are beautiful exceptions. They tend to be found on the bottom near reefs rather than on the reef like scalyheads. They also venture into shallower water, often being found under rocks at low tide.

Smoothheads have a longer, flatter snout than either scalyheads or paddeds. As the common name implies, they don't have scales on the head but they can have a lot of threadlike cirri. They have two or three cirri at the corner of the mouth, and these are often fairly thick or flattened. The scaled area is narrow, with only about 3 -5 scales in each oblique row in the anterior part. They can have a white spot at the base of the tail like a scalyhead.

Maximum length 140 mm (5.5 in).

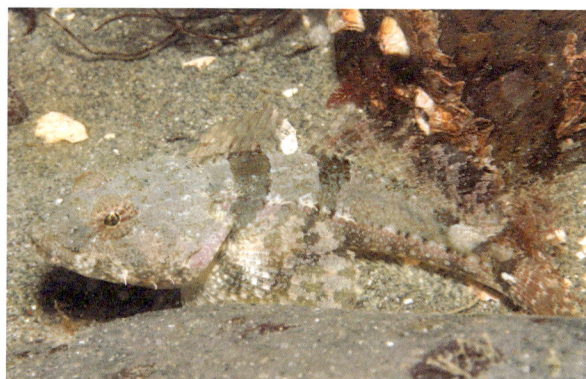

As these pictures show, they vary tremendously in the number of cirri; some even lack the cirri at the corner of the mouth. It would be worth investigating whether this is a sexual difference, or even a separate species.

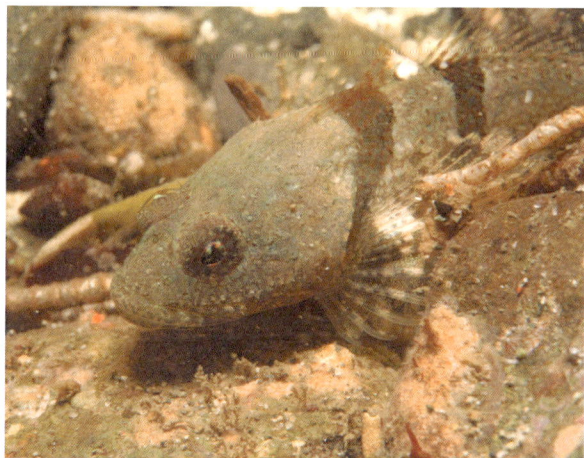

Padded sculpin, *Artedius fenestralis*

Padded sculpins never seem to achieve colors as bright as our other two species of *Artedius*. The scale band on the upper back is wider than that of a smoothhead, and the scales are much larger than those of a scalyhead (but again, this can be hard to see depending on how the light hits them). There is no large cirrus behind the eye and the nostrils and nasal spines are much larger than in the other two species; there is just one cirrus at the corner of the mouth. Paddeds tend to be on sandier bottoms than the others but are still on or close to low reefs or rocks in those areas. Maximum length 140 mm (5.5 in).

18

Dorsal view of a padded sculpin showing
the bands of large scales (left)

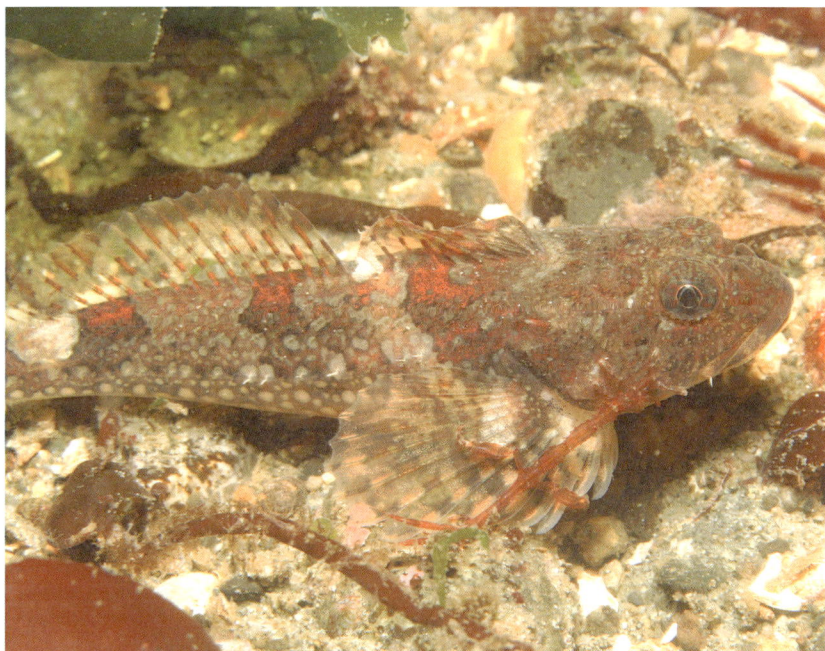

Padded sculpin with a caprellid amphipod ('skeleton
shrimp') glommed on to its cheek

Puget Sound sculpin, *Ruscarius meanyi*

This LSD was formerly in the genus *Artedius* with scalyhead, smoothhead, and padded sculpins, and is easily mistaken for one of them. It has a cluster of three fingerlike cirri behind each eye. The upper 1/5th of the eye is covered with scales, and the space between the dorsal fin and lateral line is covered in scales, rather than just having a band of them like in *Artedius*. There seems to be a lot of variation in the number of cirri on the top of the head and on the cheek, and a single cirrus may be present at the corner of the mouth.

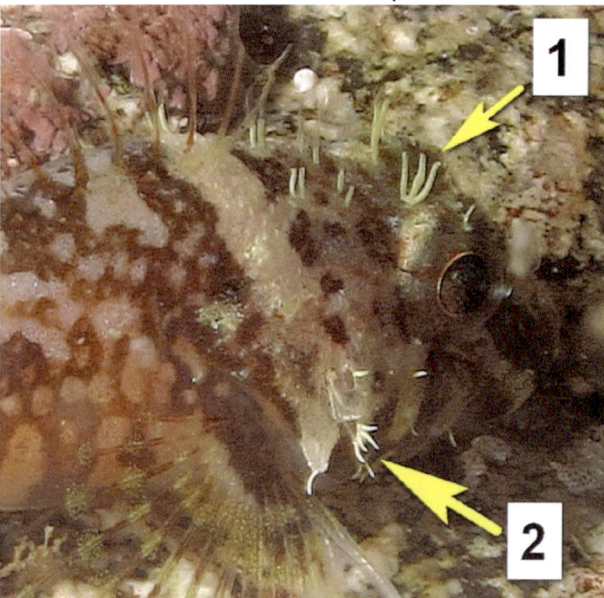

Can be confused with a northern sculpin, *Icelinus borealis*, as both have long cirri at the tip of the dorsal fin spines and large cirri behind the eyes. In northerns (and scalyheads), the cirri behind the eyes have a common "stalk" that branches near the top, while those of Puget Sound sculpins appear as three separate fingers originating from the same spot [1]. There are also some fringes of cirri in front of the pectoral fin [2]. Northerns also have a cirrus by the nasal spine that is missing in the Puget Sound sculpin.

Puget Sound sculpins are found in rocky areas and tend to hide under boulders, and have also been found inside cloud sponges. Their secretive habits probably account for their rarity in fish collections and photographs. There must be a lot of them out there, as their larvae are reportedly quite common in ichthyoplankton samples. Maximum size 59 mm (2.4 in).

Snubnose sculpin, *Orthanopias triacis*

As the common name attests, this sculpin has a very short snout. It has a broad band of scales above the lateral line, much like sculpins in the genus *Artedius* (e.g., scalyhead, padded) but the scales in snubnoses tend to be much more obvious. As with most sculpins, color varies with habitat. Maximum size 100 mm (4 in).

This species was only known from central California south, until Andy Lamb found it living at the northwest end of Vancouver Island. These pictures were taken in Monterey and San Diego.

Longfin sculpin, *Jordania zonope*

Before scuba, this little sculpin was considered rare because conventional sampling gear couldn't effectively sample the vertical rock faces of reefs where they live. With its brilliant, tropical-fish style markings it can't be mistaken for anything else in the area. Specimens living in drab areas can be fairly dull (bottom right), but they still have the characteristic pattern. Max length 150 mm (6 in).

Adult males have extra-long cirri (right), and turn nearly black during mating season (below).

Rosylip sculpin, *Ascelichthys rhodorus*

Our only sculpin that lacks pelvic fins, which makes identification easy if you have it in your hand but is of little use underwater. Still, it's a fairly distinctive-looking fish. They remind me of snailfish with their smooth, scaleless bodies and rounded fins. The only cirri are little comb-like "eyebrows" located behind the eye (lower right). The lips and first dorsal fin sometimes have reddish edges. Since they're most common in the intertidal and hide under rocks, divers should look for them at high tide in shallow water at night, when they tend to be more active. They are more common in the San Juans and Strait of Juan de Fuca than in Puget Sound. Maximum length 150 mm (6 in).

Mosshead sculpin, *Clinocottus globiceps*

This is primarily a denizen of outer coast tidepools, but does turn up occasionally in Puget Sound. With its blunt, short face, crewcut, and perpetual smile, it isn't likely to be confused with any of the other local species. One possible exception is the Calico sculpin, *Clinocottus embryum*, another primarily outer coast species that has a lot of cirri on the head and is not included for lack of a photo. Calicos have a blunt protuberance in the middle of the upper lip.

Mossheads are one of the few fish that will eat sea anemones, like this *Anthopleura*

These fish have an amazing ability to cram themselves into crevices- even between the wall and bottom of an otherwise empty bucket.
Maximum length 190 mm (7.5 in).

Unlike the other pictures in this book, these mosshead images are aquarium shots.

25

Roughback sculpin, *Chitonotus pugetensis*

This sculpin is often buried in the sand during the day, emerging at night to feed on shrimp. Although it is usually illustrated as having elongate first rays on the first dorsal fin, this is a feature that changes with age: small specimens (right) have the elongate rays, while large ones (above) do not. The large, oval-shaped eyes are also distinctive and the belly and lower sides usually have orange mottling.

I occasionally see juveniles in the daytime in shallow water. Breeding males (lower right) have bright red dorsal markings. Maximum length 230 mm (9 in).

Spotfin sculpin, *Icelinus tenuis*

Spotfins are very similar to thread-fin sculpins, but have proportionately smaller heads and don't have the long cirri by the nasal spines. The characteristic spot on the dorsal fin is only present in males.

Maximum length 159 mm (6 in).

Very small juveniles (right) don't have greatly elongated dorsal fin spines and could be confused with northern sculpins, but in northerns all the spines of the first dorsal fin are about the same length and the cirri behind the eyes are not paddle-like.

Threadfin sculpin, *Icelinus filamentosus*

This species is sometimes confused with its smaller relative, the spotfin sculpin, because both species have the first two spines of the dorsal fin greatly elongated and have a pair of large, paddle-like cirri behind the eyes. The first spine is somewhat longer than the second in spotfins and they are roughly equal in threadfins, but since they are easily damaged you shouldn't rely on it to separate them. Threadfins have a proportionately much larger and more bulbous head than spotfins, and have a cirrus at the base of the nasal spine that is lacking in spotfins.

Maximum length 270 mm (10.5 in).

Staghorn sculpin, *Leptocottus armatus*

This is the common "bullhead" so familiar to any kid fishing the shallows in our area. Though it is primarily found on sand bottoms, it turns up on cobble, mud, and shell, and even ventures up into fresh water. When handled it flares its gill covers and displays the antler-like spines that give it its common name; it can also rapidly bury itself in the sand.

Color is usually grayish with some rather indistinct saddle markings. There are no scales or cirri. This species is more skittish than most sculpins and it is much easier to approach and photograph them at night.
Maximum length 480 mm (18.9 in), but usually not over 12 in.

29

Buffalo sculpin, *Enophrys bison*

A personal favorite. Buffalos are one of the most commonly seen sculpins in the Puget Sound region, since it's probably the most abundant large sculpin on and around reefs (with scalyheads being the most common small ones). I've noticed many dive reports that mention all the 'red Irish lords' lying around are often referring to buffalos.

Buffalos can change their colors very quickly and are superb at making themselves look like a coralline algae-encrusted rock. They have a huge pair of spines on the preopercle and a series of large bony plates along the upper part of the back. The snout is blunt and steep, much more so than that of a great sculpin. Maximum length 371 mm (14.5 in).

One of the main diet items of adults is sea lettuce (*Ulva*) which is very unusual since few temperate fishes eat algae. One of my students and I did a study looking at how effectively buffalos digested this algae and found they had a higher assimilation efficiency than any reported for specialized tropical herbivorous fishes.

Red Irish Lord, *Hemilepidotus hemilepidotus*

When it comes to changing color to blend in, no Northwest sculpin can match the skills of a red Irish lord. I suspect the speckling so commonly seen on the eyes of these fish is a means to camouflage it, since an undisguised eye can give away the location of even the best concealed fish.

Irish lords have a distinctive band of scales that loop around front of the first dorsal fin (below). They are also one of the few with a large flattened cirrus at the corner of the mouth; most species have threadlike cirri. The eyes are unusually large.

Maximum length 510 mm (20 in).

This Irish lord has not only perfectly matched the color of the coralline algae, it has also matched the speckling caused by small tubeworms in the algae.

An example of an extremely drab red Irish lord, living in an extremely drab area.

Brown Irish Lord, *Hemilepidotus spinosus*

Drably colored red Irish lords (like the one at the bottom of the previous page) are often mistaken for the closely related brown Irish lord. Browns are an outer coast species that is rarely encountered. I found the one shown at Sekiu, Washington; it was the first that I have positively identified and it was too far back in a cave to get a full body photograph.

The two species differ in the number of scales, both in the dorsal rows that wrap around the first dorsal fin and the scales below the lateral line. The upper scale band of red Irish lords is 4-5 scales wide; the ventral band is 6-7 scales wide. The numbers are essentially flipped (6-8 dorsal; 4-5 ventral) in browns. Scales are probably the only reliable way to tell the two species apart, so if in doubt always get a shot that shows the dorsal scale row, which is easier to see and count.
Even if you can't get an actual scale count, the relative size of the scales is different and can help in telling the two species apart. Reds have noticeably larger scales, because the width of the bands are about the same.
Brown Irish lords are smaller than reds, topping out at 330 mm (13 in).

Great sculpin, *Myoxocephalus polyacanthocephalus*

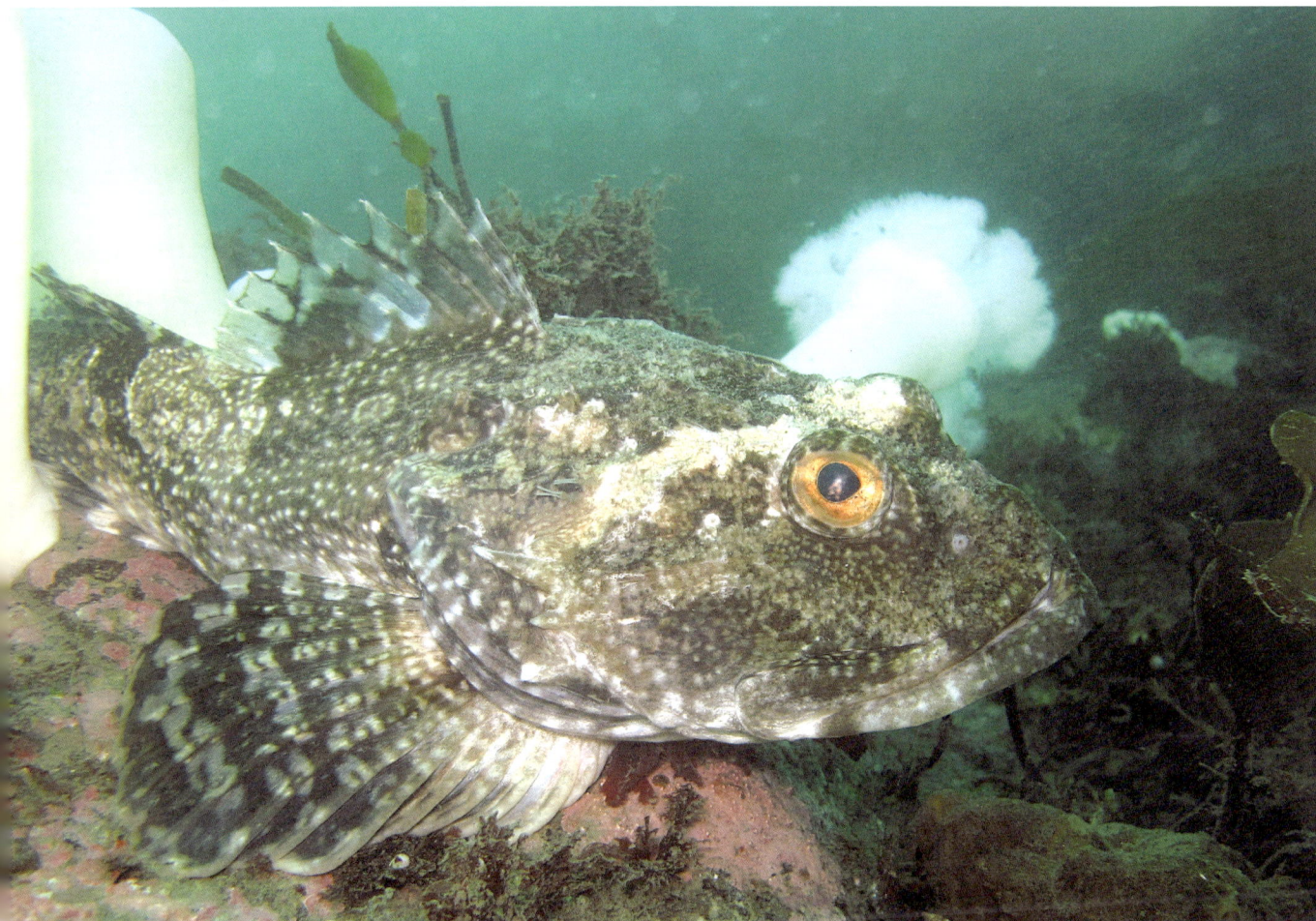

This is the second largest of our sculpins, who tries to compensate for second place with the longest scientific name. Great sculpins have an oversized head (having swelled from being called "great" so much?) and a sharply tapered body. They have a median ridge on their snout that can be seen in profile and the undersides are usually strongly spotted. In adults the rear of the jaw extends back well past the rear of the eye; in juveniles it may only reach the rear of the pupil.

adult

juvenile

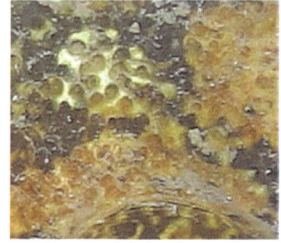

rough, 'pebbled' texture

Great sculpins turn up in all sorts of habitats, from rocky reefs to eelgrass beds, and some of the largest specimens I've ever encountered have been in eelgrass. Of the eight species of *Myoxocephalus* that occur in Alaska, only the great sculpin occurs as far south as Washington. Maximum length 760 mm (30 in).

Cabezon, *Scorpaenichthys marmoratus*

The other common name for this fish is "giant marbled sculpin", which aptly describes both their distinctive color pattern and their ranking as our largest sculpin. They can reach about a meter in length, but it's their girth that makes them truly impressive. Cabezon feed primarily on crabs.

Like many other sculpins, cabezon have a pair of branching cirri behind the eyes. They are unique in having a single flaplike cirrus near the tip of their snout, and this is a good character for recognizing the juveniles which look different from adults in color, pattern and form. This cirrus seems to mostly disappear in large specimens. (Grunt sculpins often have a single cirrus on the snout, but are different in so many ways the two could never be confused.) Maximum length 990 mm (39 in).

Male cabezon vigorously defend their eggs, which are toxic to eat. In other words, if daddy doesn't get you, the eggs will.

Slim Sculpin, *Radulinus asprellus*

Slim sculpins are perhaps the sculpin most likely to be mistaken for a poacher. They have huge eyes, a very distinctive long, sharp nasal spine, and three very widely spaced dark saddles. Watch for them on sandy or muddy bottoms at night.
 Maximum length 152 mm (6 in).

Ribbed sculpin, *Triglops pingeli*

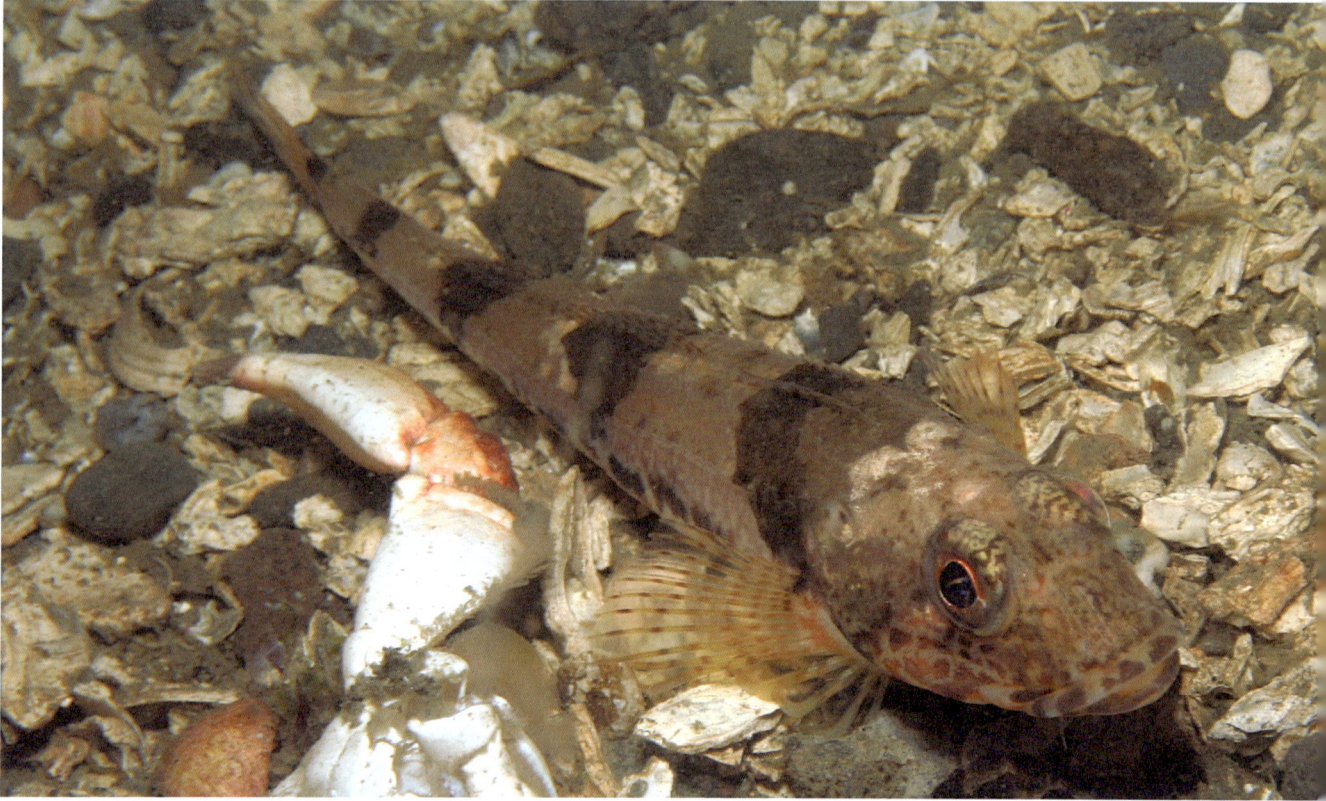

The two species of *Triglops* in our area always remind me of poachers due to the large, platelike scales along the lateral line and their stiff, straight posture. Both have oblique folds along their sides that, depending on the lighting, may not show up in a photo. If you can see it, the best way to tell the two apart is by the pectoral fins: in roughspines, some of the rays of the pectoral fin are longer than the rest and fingerlike, without webbing on the outer half. In ribbed sculpins all of the rays are fully webbed. Although both species are fairly long and slender, ribbed sculpins have a noticeably thicker build than roughspines. Male ribbed sculpins have a dark line down each side, at least during mating season. Look for ribbed sculpins on open sand or mud bottoms. Maximum length 232 mm (9 in).

Roughspine sculpin, *Triglops macellus*

As noted under the ribbed sculpin entry, these have a distinctive pectoral fin with long, fingerlike rays in the lower part. Like ribbed, they have oblique folds along their sides below the lateral line, but they have a much skinnier body than that of their relative. They are typically found at night on open sand bottoms.

Maximum length 300 mm (11.8 in).

Grunt sculpin, *Rhamphocottus richardsoni*

A serious contender for world's cutest fish, this species is so distinctive it doesn't require a description. Though usually thought of as a reef fish, they're one of those species that can turn up on any type of bottom. Grunts are now in their own family, the Rhamphocottidae, and do not appear to be closely related to other sculpins.
Maximum length 89 mm (3.5 in).

It's been suggested that grunts 'mimic' giant barnacles because when they use empty ones for shelter, their head shape has some resemblance to a live barnacle. I think it's just a coincidence, as there are many other fish that also use these neat little homes and there is no evidence that grunts specialize in this kind of habitat. Sometimes a cigar is just a cigar.

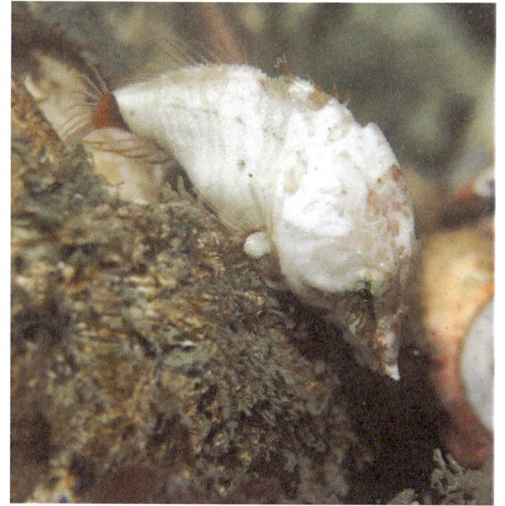

Small juveniles are sometimes nearly all white like this one, and occasionally this color pattern persists in adults

Silverspotted sculpin, *Blepsias cirrhosus*

Silverspotted sculpins never fit the usual image of a sculpin, so t's fitting that they're now in the more swanky Hemitripteridae, along with sailfin sculpins. In addition to being unusually fancy and frilly, these fish are also very active swimmers as they hunt for amphipods in algae and the kelp canopy. Color is usually dark brown with silver patches on the side and an orange belly. There are three pairs of long cirri on the lower jaw.

Maximum length 200 mm (8 in).

Sailfin sculpin, *Nautichthys oculofaciatus*

This is one of our most distinctive fish, with its tall first dorsal fin and undulating second dorsal fin. Sailfins are nocturnal shrimp predators. Even in the daytime, crammed into a crevice with all their fins folded down, they are still easily recognized by the diagonal black stripe that runs from the cirrus on the eye and down across the cheek. The first dorsal fin varies a lot in shape and height and can be quite frilly in juveniles. Maximum length 203 mm (8 in).

Small juvenile

Tadpole sculpin, *Psychrolutes paradoxus*

This is a fairly common little sculpin on open bottoms that have eelgrass or macroalgae. Although they don't seem to hide in the daytime, they're easier to spot at night when they tend to sit out on the surface of algae blades. The one in the bottom right photo is eating a cumacean. They have blobby, tadpole-shaped bodies with white markings. The species they are most likely to be confused with is the soft sculpin, *Psychrolutes sigalutes*. It is not included as I do not yet have a photo, but a good one can be seen here:
https://nwdiveclub.com/viewtopic.php?f=11&t=5931
Maximum recorded length for tadpole sculpins is 65 mm (2.5 in).

I've found that these small fish are unpalatable to most predators, which probably explains why they are able to sit out in the open with impunity. Larger sculpins will "cough" and spit them out unharmed shortly after swallowing them, and rockfish spit them out instantly.

Blackfin sculpin, *Malacocottus kincaidi*

Blackfins are rather flabby sculpins with enormous heads. The pectoral fin is light in the middle with a dark margin and the tail has dark bands; there are no spines or bumps on the top of the head. Juveniles (below) are almost as cute as a lumpsucker. Maximum length 106 mm (4.2 in).

The spinyhead sculpin (*Dasycottus setiger*) has a similar body shape, but numerous cirri and some large knobs on the top of the head. Like blackfins, they are rarely-encountered deep water fish. See good photos of one here:
https://nwdiveclub.com/viewtopic.php?f=11&t=19098